Peter J M Clive

The end of the age of fire

Contents

Prediction 1
Dies irae 2
The rowan berries have turned black 4
Amazon 5
Asteroid 8
Astronaut 10
August 2020 13
Climate change 15
Rex quondam et futurus 17
Aestivation 18
Solstice 21
Chimpanzee 29
Royal baby 30
The ark 31
Leviathan 34
People don't listen 38
Lifeline 39
The end of the age of fire 40
Scream 60
Shout 63
Kingdom come 64
The trials 65
The eighteenth green 66
Time 84
Tired 86
14 87

Sloth	88
Jump cut	90
The well	93
Eden	95
Decade	97
For Siddhu	100
Ashes	101
Can of worms	102
The fate of the poets	103

Prediction

We live in the era

of the weary Cassandra,

the jilted Jeremiah:

all prophecy exhausted,

the oracle is hoarse,

too tired to testify,

witless with witness fatigue.

You heard the warnings.

You chose to ignore them.

It's your turn to scream.

Dies irae

Our lives are increments
by which the world changes
from a garden to a grave,
the sexton's spade a plough
turning the soil since dawn in Eden
to the end of days, and He
who first enclosed this acre,
has returned, as was foretold,
to decide what now becomes of us.

A fanfare announces His arrival,
scattering sounds like seeds
to root through every tomb in the land,
and in their soil find hidden sin
to bring into bloom. What graves
will the Gardener find rank with weeds,
to scythe into the fire? What fragrance
will our souls yield for Him to cherish
and preserve? Or will He be too late?

Impatient for judgment,
we have judged ourselves.
Drought and deluge, fire and flood,
famine and strife have left an arid Eden.
There are no more graves.

The stone of every tomb crumbles to dust,
and is swept and stirred by hot siroccos
into anonymous drifts, heaping in dunes
our indistinguishable delinquencies,
our inextricable souls beyond salvation.

The rowan berries have turned black

The rowan berries have turned black.
The birds have not been back
to branches they last year stripped bare
as they have done year after year.
A cursèd crop hangs there.

Its blush has succumbed to a blight
in times that lack sense and foresight,
tomorrows scorched by yesterdays
when we unleashed the flood and flame
so torrents rush and wildfires blaze
but none are found to take the blame
for decades of delays, delays, delays,

and so our appetites unchecked
mean little birds' must now neglect
the rowan berries that are black.
The birds will not be back.

Amazon

The air is thick with lies
used to sell us things,
and the world burns.

We can't see clearly
through the fog of adverts.
We can't see the wood for the trees,
We can't see the should for the need,
the spurious must haves that deceive,
and the world burns.

We take our axe to Yggdrasil
for a thirst no rain can quench.
Oil and water separate
into plastic bottles
and designer effervescence,
and the world burns.

Our response to the crisis
is posture and affectation.
Deadlines are set, promises made,
all while the world burns,

The world burns, and we are told
we want a better fire,
we want everything the fire provides,
but more, newer, bigger, improved,
sexier, shinier, elite, exclusive,
and the words that we could use
to contradict this lie
have been stolen by the arsonists.
It is too late. We no longer care.
All we want is stimulus and reward.
We are enslaved by freedom,
bought and sold
by the choices we are given

until all we can do is agree,
and immolate ourselves
for their entertainment,
douse ourselves with petrol
pulled from the remains
of long dead creatures,
squeezed from the Earth like an orange,
sucked from the marrow of strata
cracked open and exposed
to a storm of greed,

and once everything is soaked,

it is ignited with a spark

of renegade desire

kindled by private shame

whispered to our secret selves

by the false digital priests

to whom we confessed our appetites,

and the world burns,

and the empires that arise

from the ashes of our ecocide

are death cult fantasies, dancing

through the twisting wildfire smog

to seduce the void

while the world burns.

Asteroid

The sun and rain clouds share the same sky today.
Wet slate gleams like steel. The sound of heels
hurrying on a pavement. Showers and gusts
chase pedestrians, but what do I care
about the sudden panic of their umbrellas?
The asteroid hit five years ago.

The slow dislocation of the seasons
escapes their everyday attention,
but it doesn't matter, it's already too late.
They don't need to worry about catching up.
They have already been overtaken by events.
The future is already out of reach.
The steps to checkmate dwindle
in an endgame they don't know they play.
The asteroid hit five years ago today,

and all these people act like nothing has happened,
care about all the little things I never got to care about.
Ridicule and ostracism and loneliness
blighted my monstrous youth and preoccupy me now
even as I walk among them,

and so, even though they are all doomed,
I don't care anymore. They are the asteroid,
and the asteroid hit forty years ago,
and it's been too late for far too long.

Did I invite the impact? Did I ask it
to come crack open the permafrost,
unlock millennia of methane
and introduce them to what it's like
to be estranged from the very air you breathe,
to be an alien on a planet that may as well be dead,
to teach them the utter pointlessness of my existence?
That would make sense, except
the world does not deserve to suffer.

There is no judgement. God is impotent
before those who condemn themselves.

Astronaut

Who knew that the fragility of the astronaut,
alone, far out in space, riding his leaky kettle,
whistling through the airless void, his panic
attenuated to such a monotony of fear
that he has turned the bag he must use
to breathe in and out, slowly, steadily,
to control his breathing and calm himself down,
into an entire suit, with a helmet, that he wears
the whole time he is in his capsule
cobbled together from the best bits of junk
the world has to offer,
who knew that his fragility was the rule
rather than the exception,
and was only the most extreme example
of an everyday and unremarkable heroism
we all share as we sip tea made with water
strained through eroding soils and receding roots,
our kettle a climate that is out of control.

Who knew that the point of the Moonshot
would turn out, not to be to land a man on the Moon,
but to establish the rest of us safely here on Earth,
if we have the "right stuff" for the challenges we face.

Listen to those who do not know,

who refuse to know or understand, listen,

no matter how tedious or tiring it is,

not because they say anything worth listening to,

but because others listen to what they have to say.

Their gospel of wealth without responsibility,

the simple, stunted arithmetic

of their calculations without compassion,

the centrifugal trajectories to fantasy destinations,

the society shredding contradictions of their priorities:

this is how we ended up where we are,

in the wrong place precisely when it matters most.

Listen the way you listen to a fool who has stumbled

into responsibilities beyond their understanding.

Listen the way you would to talk someone down

who thinks they have no other way out.

If you ignore them all that happens is

their point of view persists

until it overwhelms and destroys us all

in the wreckage of a mission whose failure is so total

it leaves no-one to write the letters of condolence,

no family to receive them, no luxury of tears,

no time to grieve, just the void that we confronted

and from which we learned nothing.

And once you've listened, reply.

Tell them instead that we are all astronauts,

embarked upon the most exceptional adventure,

on a planet orbiting a star that orbits a galaxy,

one among a vast congregation of other galaxies,

that the universe is more wonderful than we can imagine,

and that beauty is proof of the miracle of survival

when our imagination fails to encompass it,

and it will take every ounce of effort

and every glimmer of wit

to land ourselves safely back on Earth

after the journey we are on.

August 2020

We make artefacts now,
not things to use today
but things to be found
in an archaeological dig
once we are all long dead,
so that people can marvel
at what we were capable of
given we were all mindless thugs.

We fashion jewellery now
so we can swallow it
with the other valuables
in the hope someone finds it
among our anonymous ashes
in some future excavation
and we are remembered
in a museum label at least,

and if we are seen, as we polish
some precious stone upon our sleeve
we are told that we are mad,
and the gem is seized and shattered,
and ground down to form mortar
to bind the unremarkable bricks
with which they build the walls

that surround us now, within which

our remains may one day be discovered.

Climate change

Your words make my brain disintegrate.
I can't keep up with your lies. Logic is mush.
You prove things wrong by saying they are wrong.
According to you things are right
because everyone knows they are right.
I have neither the time nor the inclination
to keep up with your incessant deceit.

I see you, 100,000 years ago. I remember you.
I remember you. I was there when you said this before,
100,000 years ago, and we ignored you then too.
You hate that we ignored you.
You think it was because we hated you,
but we don't hate you. Up your game!
You've waited this whole time,
from when our mother made the first needle
and sewed the clothes everyone has worn ever since
without a single nod of recognition from the catwalk,
you've waited while bigger scandals broke,
and let women and snakes take the blame,
and when you had the chance to make your move
you waited, and you choose now to speak?

Rewind, mate. You've learned nothing.

Do you think that because we let you have you say,

anyone thinks you have anything worth saying?

You killed the beast that didn't need to be killed.

You sheared the sheep whose wool we didn't need.

You give us what we do not lack

and leave us without what we most need.

You prove the point that doesn't need proving.

We are not going to take you seriously

just because you can now threaten to destroy the world

with your vanity. Kill us all. You'll still be wrong.

Wear this world like a bead on your necklace.

You will not escape our ridicule from beyond the grave.

Rex quondam et futurus

I wear no crown.

I hold no throne.

I lead no army:
I am unarmed.

I have no honour,
bear no grudge,
and claim no allegiance.

All those things
that I once had
passed long ago
so I may be again
that which I always was,

and silently my summons
gathers bird and beast, responding
by instinct, not obedience. A profusion
of apple blossom is my banner, surrendered
like ten thousand white flags, scattered
upon the waters, from which the weapon
that I once discarded is restored again,
bright and keen, into your hand.

Aestivation

Hibernating in summer,
waiting for it to be over:
withdrawing from a world
no longer rotating on its axis,
a world that has twisted its ankle
and stumbles, so the seasons spill
their heat too soon, or late,
drought competes with downpour,
and time is a river run dry
or a torrent in spate,
and men form glaciers of hate
that advance down high street
and into marketplace to guard statues,
themselves dumb sentries of the state,
as a hot ice age of anger rages;
so I withdraw from summer pastures
and peonies and skylarks to my cave
and wait for this to be all over.

Remember the taste of stolen food?
Remember, as a child, eating apples
purloined from radiant orchards
ablaze with the tongues of sparrows,
orchards whose walls we scaled as easily
as fortifications patrolled by toy soldiers
in our rooms on rainy indoor afternoons;
cobs twisted from thickets of corn
whispering and swaying mysteriously
in some summer breeze only they felt;
carrots plucked from soil near fences
through which we reached
at the edge of the field;
and childhood the secret passport
to all these ecstasies of transgression
pursued together in the primal loyalty
of common purpose, or in pristine solitude,
in the endless summer days of childhood
that had two flavours, the mother's love
that was always in season, and the taste
of stolen food in summer? Dream of that.

Dream of that during this summer sleep,
dream of that in the unremembered dreams
that stir deep in your summer slumber.
Dream of well-fed and unworried days
before the vexations and the drought
of this uncommon age. Dream of that,
when to awaken is to weep,
and carry that dream through the dark,
keep it safe, and one day, I promise,
the pessimism of the mind will yield
to the optimism of the heart,
and we will break bread together
and eat, when this is all over.

Solstice

We collected our ancestors remains from the cairn,

bundles of bones wrapped with reverential hush,

and conveyed them with all due ceremony

and gathered with them at the standing stones

to watch the dwindling of days,

the sun rise later and set earlier than before,

the gap between them growing more narrow

as the sagging arc traced by the limpening sun

sank further past the darkening world's edge

and we wondered: will the light continue to wane

until we are deprived of it entirely?

Will the pattern our ancestors saw

reliably enough to raise these stones

as a record of the seasons of their days

save us and restore the light? If not,

how can we begin to repair the heavens?

Should we adjust stones that have stood

for one hundred generations, to avert the end of days?

Having received no promise that we could understand,

we rejoiced with no sense of redemption

when some celestial mechanism delivered the sun

a little earlier one morning than the day before.

We saw as it retreated from the extreme on the skyline,

the mark on the world's rim it reached every year

among the silhouette of the trees'

cold bare scaffold of the next year's seasons,

just as we had hoped and prayed and come to expect.

We still feared one day it would not find its way

back to us from the depths of winter.

The darkness remained, as a threat,

a prototype of evil, arbitrary, evasive,

prowling around the plots of light

our campfires placed within the deeper night,

destructive and inscrutable, unamenable to any bargain,

consuming worlds not out of hunger,

but out of boredom and caprice

in an age before we invented the devil

to take the blame. Nothing happened for any reason

of the kind we subsequently came to understand.

The palpitation of the seasons was ensnared at last.
Within the shaman's feverish delirium,
at the back of the cave, from his trance,
the first truths were received,
truths we could hold and share and use
to attenuate and curtail the terror of the dark.
Insights extracted from his ravings, narratives
teased like thread from simultaneous experience
tangled up in hypoxic convulsions in the sacred smoke
became our campfire tales. Zodiacal heroes
endowed with superhuman strength, bearing divine gifts,
prevailed in elaborate contests set by whimsical gods,
victories over champion monstrosities
whelped in a cave by the mother of monsters,
and so we began to feel reassured,
and slept the first human sleep,
enjoying at last depths of slumber unknown to any animal
still nervous and tormented by the constant threat
which we now eluded and began to forget.

Within the safety afforded to the precincts of our dreams

we built shining cities, cascades of glass, tumbling down

through the sparkling cataracts of empire,

falling over the precipice of history, and rising again

behind the hooves of each departing horde,

and in the glistening procession of millennia

some among us at last discerned the workings of the world.

Truth swung so low in her orbit

her necklace of planets hung within the astronomer's grasp,

and were plucked and strung like pearls

along the sinews of their reasoning.

Musicians sought solitude in the forest.

They found that truth strayed so tame across the forest tracks,

lingered so long within earshot

without being startled by their presence

that her song was overheard and learned by heart.

The dark at last was shrunk to fit the sleeve of science.

Our fear of it softened to a surly scowl

glimpsed between the pillars of light

within which it was confined.

But the dark was not defeated.

Within our marrow the memory of our fear lingered,

and now a different kind of solstice comes.

The fulfilment of expectation has grown so tedious

that any other sequence of events is inconceivable,

but the light continues to spiral down night's dark drain

past the level from which we are accustomed to see it fill again.

Generations of certainty suddenly seem naive and idealistic.

The natural order is suspended. The solstice swells and swirls

and becomes an irresistible pole to which we are pulled

to be devoured by the world's strident new solipsisms.

Our ancestors' remains are scattered. We revere instead

those who are briefly notorious and quickly forgotten.

Indolence becomes our tutor and scorn his lesson.

"Darkness is good," advisers tell the powerful.

Their project is to stall the solstice. The point

around which the year is slung back out of shadow,

the celestial citadel where the year is rescued,

itself is captured. The sun is taken hostage. Darkness reigns.

Time becomes indefinite, adrift. The solstice continues

well past any familiar reckoning. Darkness persists.

Long forgotten fears stiffen us again with sudden paralysis

in the extended contemporary solstice of the mind.

Accuracy is irrelevant. Tribe matters.

Information is discarded or disregarded.

The confirmation of any bias is preferred,

and if you can claim to have seen through it all long ago

your reward is a place around the totems that have replaced truth.

We no longer rely on such effete, decadent contrivances as reason.

We grow strong in the dark, unseen. We do not need to be consistent.

We do not find fascination in truth, only in darkness.

Alliances are temporary expedients rather than alignments of ideas.

Allies are discarded as soon as they cease to be useful.

Statements resemble rather than convey knowledge.

We use them as a way of belonging rather than knowing.

We do not need to describe the world, only our place in it,

with Shibboleths, passwords, noises like the call of an animal.

Conspiracy replaces truth as a totem we can gather round,

and if this is the only purpose our words serve

they are truth-proof, for they lose none of their force

no matter how often they are contradicted.

We distil the world into competing visions,

and discard the evidence for any of them,

and this is how we imagine we will be redeemed.

Now we gather around our own isolated campfires,

warming our hands against the burning books,

telling tales of how the sun was just a trick,

a ploy our enemies used to enslave us long ago,

and how the darkness has set us free,

and will endure for a thousand years

or whatever the highest number is

that anyone can still count to

before being cast out as one of those experts

while our enemies sit around their own fires

telling the same tale. Huddled around fires

we whisper rumours of distant wars,

unimaginable evils stacked up

and propped against each other

on the endless shelves of imagined history

where we condemn each other,

until our accusations embrace

like exhausted boxers,

while in the dark,

beyond the light of the campfire,

beyond the flickering familiarities

to which we cling in its unreliable light,

the solstice prowls in the dark once more,

its sable pelt velvet and invisible,

unknowable,

waiting to consume us.

Chimpanzee

There must have been a quarrel,
long ago, so very long ago
no-one remembers the cause,
but bitter enough to last this long,
past all chance of reconciliation,
it seems, because there you sit,
on the other side of the glass, in this zoo,
at your end of the seven million years
of our separate ways, like me,
wondering what on Earth happened.

Royal baby

If only penguins were royal babies.

If only whales were royal babies.

If only insects were royal babies.

If only birds were royal babies.

If only we were royal babies.

If only everything as rare

and beautiful and precious and endangered

as a royal baby was a royal baby

If only the Earth was a royal baby.

The ark

God said unto Noah:
"Of every clean beast
thou shalt take to thee by sevens,
the male and his female:
and of beasts that are not clean
by two, the male and his female."

And there went in two and two
unto Noah into the ark,
the male and the female,
as God had commanded.
They went in unto Noah into the ark,
two and two of all flesh,
wherein is the breath of life.

Except this new ark made by man,
this latter-day ark of extinctions
which we have prepared for the beasts,
cannot contain the breath of life.

The animals now retreat back to the ark
that saved them once, their numbers dwindling,
all abundance and fruitfulness depleted,
until only two of each remain, walking backwards,
rewinding themselves into the ark,

except this is a new ark, made of steel,
a tin pot ark, not a product of necessity,
an ark that makes its own emergency,
its dimensions not reckoned in cubits
but on balance sheets and annual statements,
an ark of many mansions,
in which each beast may find its place:
an abattoir built on land we cleared of forest
where we once innocently hunted;
a syringe another, to inseminate livestock
in a parody of what they once did two by two
before we divided those animals we could tame
from those we couldn't, with extinction,

an ark of extinctions that we have made,
not because of any divine command,
not to escape a fate ordained for others,
but to go back and join those others,
to exercise our free will and defy God,
to insult Him by rejecting our salvation,
to soil the rainbow with air pollution
and tell Him we don't care about his promise,
to tell him we are going to go back
to join those who drowned,

we are going to summon the waters ourselves,

from every glacier,

from behind every rampart of ice

raised against the sea in Greenland and Antartica,

we are going to change the climate,

and we will get back into the ark

and go back to perish with the wicked

because God made a mistake saving us,

and it's a mistake we are going to fix.

Leviathan

Madness begets madness.

Our indifference towards those in genuine need
is matched only by our own fantasies of victimhood.
We complain when short-changed then walk past beggars
as we reflect on our own petty vindications. On holiday
we complain about slow service at the same beach resort
where bodies of drowned refugees washed up the year before.

Madness begets madness,

and even to scrutinise our condition invites ridicule and scorn.
Argument is taken as affront, evidence dismissed as provocation,
logic taken as personal insult. There is only shared conviction
or deliberate contradiction. There is no nuance, doubt or investigation.
Assertion is an end in itself. Facts are banners around which to rally.
News is fake unless it makes us feel good. Lies are celebrated
if they let us to continue to lie to ourselves. Hate and avarice prevail
and the dog-whistle and the knee-jerk are our only clarion and litmus.
Truth corrodes and crumbles in the acid and greasy bile
with which we lubricate our discourse. We succumb
to the suggestions of every opportunist, indulge every easy prejudice,
genuflect to monsters, and replace thought with reflex and gratification,

and madness begets madness.

We must put our hands over our children's ears
when our leaders speak, to protect them from the filth they spout
as they exculpate themselves from their sex crimes, implicate all men
to make their own guilt seem commonplace and unremarkable,
misdirect our indignation towards the weak
and the vulnerable and the foreign.
We placate the angry at the expense of moderation.
We indulge the bullies who insist we listen to them
despite the fact they have nothing to say,
who complain they are being ignored
even though no-one else is allowed to speak.
We allow others who timidly bear the most urgent of messages
to fall silent, to appease them and let their loud, empty words
fill our halls and chambers and courts and parliaments
and cathedrals with their specious import.

But do not confront the world head on. If you do
you will only discover it possesses inexhaustible reserves of horror
that will overwhelm the best in anyone.
It will outdo even the most inventive imagination
with the variety of its atrocities.
It leaves God dumbstruck and paralysed with trauma,
unable to intervene, or so it seems.
What chance do we then stand? Is there is another way?

Even though we try, even though we may raise a humble yard
from the broken limbs and timbers of our greatest cities,
until its pillars become our tallest skyscrapers once again, its canopy
our hand held up against the sky as the measure of our ambition,
the highest degree of civilisation we can accomplish
remains an exquisitely refined barbarism. We may invest
the finest vessel we can construct with the noblest cargo of hope,
and launch it with our age's loudest boast of progress,
raised in unison as a cheer to hail the brightest future,
but Leviathan will splinter its stiffest beam in its grip,
and reduce all collective endeavour to disconnected tales
of individual panic and despair,
and to survive we must become flotsam and seek that beach
where we intrude on another man's dream of paradise,

but accept the futility of our efforts and make them anyway.
The most we can ever achieve is temporary repair,
so defiantly live in a world built and rebuilt from its own wreckage,
and act not out of fear or desire for the consequences of our actions,
but because our actions are who we are
and we will not be defined by their futility,

and be kind to one another. Offer the beggar change.
Learn the histories of the places you visit,
reflect on the losses they have suffered,
celebrate the human legacies they have left us.
We will dismantle and fold up and pack away
and dispatch paradise to people in distant lands
going through hell, so that some earthly shelter
may be assembled where it is needed.

For kindness begets kindness.

People don't listen

and when they listen,
they don't hear,
and when they hear,
they don't learn,
and when they learn,
they don't care,
and when they care,
it's too late,
so they don't listen.

Lifeline

Think in terms of lifelines.

Don't think in terms of deadlines.

The deadline isn't 2050 or 2030

or even right now.

The deadline was 5 years ago.

We missed the deadline.

We don't need a deadline anymore.

We're dead already.

We need a lifeline.

Think in terms of lifelines.

The end of the age of fire

I.

We were grave robbers,
disinterring the dead of all past ages,
things that drew their last breath
long before we drew our first,
splintering flat black strata
where ancient forests once lay down,
siphoning ancient darkened sunlight
from the veins of the dead,
where it had congealed and pooled,
clogging subterranean arteries.
Our hearths were their pyres,
where we cremated geology,
in whose heat and light
we raised forty thousand generations,
taught them to stand up straight and walk,
inducted them into the ways of fire and thread,
the knap of flint, the hunt, the toil and trade,
inducted them into the age of fire.

II.

We were grave robbers and we were orphans.

We were orphans of the age of fire,

abandoned by the Earth, so we were told,

even while we mounted her breasts like gluttonous infants,

even while we flattened and augmented our hands,

to make shovels with which to delve and burrow

through the folds of her moist black flesh

tumbling over us in dark cascades of abundance,

yielding to our ploughs where we pulled them,

and offering us bouquets of wildflowers where fallow.

III.

We were orphans, abandoned by the Earth,

so we were told,

so that her gifts to us would seem like theft,

something taken, not given,

a lie to legitimise all subsequent cupidity,

and we invented scarcity in a world of plenty,

controlling access to abundance,

to create imaginary advantages

from which to conjure the rights and privileges of kings.

We didn't see her.

We didn't know she was there,

even though she stood right in front of us.

We imagined ourselves to be orphans

even while our mother kissed our forehead

with gentle dawn mists rising in the morning

and beams of moonlight glimpsed through clouds at night,

which we could not see

because our minds were filled with smoke and fire.

IV.

Our fathers said we fell from the sky.
We are the sons of gods. We are those golden sprues
that snapped off the Sun's crown and fell to Earth.
They said we are all stardust.
Our fathers' faith forgot the mother
who swept that dust into a heap
and brought the rain that made the mud
from which we were formed.
Even her memory was demonised:
Lilith, Leviathan, Lamia, Echidna, Tiamat,
Medusa, harpy, banshee, succubus, siren.
We lost all sense of legitimacy,
and lived as exiled bastards, wandering nomads,
even while we rested in her embrace.
Our thrones, that she had prepared and sanctified,
that anyone who knows her can occupy,
stood empty and abandoned
and overgrown among the roots and limbs
of the sacred groves we now cut down
to feed the fire.

V.

The sacred precinct was overthrown and the fire was refined.
Its uses extended beyond the heat and light of the campfire.
It became a beacon, a torch,
a mighty Pharos raised to guide ships to harbour
across tumultuous seas raised against us.
Eventually the light of that beacon fell across the mind,
and all that was understood was seen in that light,
and comprehended in terms of the commerce it enabled,
and the fire was honed until it became hot
and narrow enough to cut through metal,
great metal sheets with which we clad the ships of our mind
as they traversed new realms of knowledge,
welded fast and watertight.
The electricity and magnetism of Faraday in the hands of Maxwell
described the light at last,
a gospel interpreted and preached by Heaviside
until Einstein took it from him to make light itself the hinge
on which swing time and space
with the weightless cosmic telegraphy
that carries the whisper of time from place to place.

VI.

We were orphans and we drank, not milk, but fire.
We hauled slick black cables of coagulated blood
from the ancient arteries of the earth
and boiled it in our towers to restore it to life
- at least, that one single aspect of its former life
that we can use: it burns.
We burn clotted scabs and crusts of coal,
and a million years of fire ensues
over the forty thousand generations of our species
adrift and alone in our metal ark upon a sea of fire.

VII.

We burned effigies of the mother

we were told had died and abandoned us

with the mortality we inflicted on her fruit.

We attributed the fate of what we had killed

to the mother that had made us

to exculpate ourselves of its murder.

We punished her for that death

with which we believed she had abandoned us.

We called her harlot. We called her witch.

The tenderness of her incantations

was lost on those who did not see the world

the way her serenades described it,

for whom the life she celebrates

is a skeleton circus of the imagination.

We explored her with all the intricate contrivances of torture,

telling ourselves the atrocity was autopsy,

telling ourselves we were motivated by science, not selfishness.

We dismissed her screams as the residual reflex of a corpse.

We performed every unwitting violation

and the world turned sour as it stood.

VIII.

We cursed our thwarted appetites even amid plenty,
invented entitlements to allow us to ration abundance
and defended the inevitable injustices of our grim calculus
at the barricades of revolution against rival reckonings.

IX.

The fire vouched for every other appetite

with the implacable voraciousness of its own

and so we thought it reasonable to consume each other

as indiscriminately as we consumed the world about us,

with no regard for what we damaged or destroyed along the way,

and so our society became a process of digestion

that reduced all virtue to the price it can fetch at market,

and we were taught to tolerate this betrayal by betraying others,

that this is the way of things in the age of fire,

whose principle is scarcity rather than abundance,

as the fire consumes without replenishing its source

and we emulate the fire in our greed, our recklessness,

our disregard for the earth.

X.

We thought she was dead because we burned her,
we didn't burn her because we thought she was dead.
We didn't burn women because we believed they were witches,
we believed they were witches because we burned them,
and the piles of dead livestock burned to eliminate contagion,
the beacon of Grenfell shining through the small hours,
the artificial sun blossoming over Hiroshima,
the chimneys of Auschwitz,
all testify to the upside-down logic of the age of fire,
and filth spills in plumes from the bonfire
and mixes with and marbles our clouds
and bestows on us the most magnificent of sunsets
with the ash of that artificial volcano that is our civilisation
in the days before the world's end,
the days before the end of the age of fire,
and when the end came, as it must
those whose job it was to keep the fire burning,
in a fit of petulance, threatened to burn everything down
and turned the fire into a weapon, and war was born,
making real orphans among the rubble of bombed cities.
We'd know more, but they burned the evidence.

XI.

But there was no fire. There never was.
The destruction was always entirely our fault.
There was a congregation of hot incandescent vapours
in which we chose to see a fire
in the same way we choose to see a blade
hidden in a lump of virgin flint,
see the shaft of a spear in every pine,
see a profit withheld from us by the savage
until he makes the trade that is our manifest destiny
at spear-point if needs be: we'll burn his village if he doesn't.
The fire was always just an excuse. We were the fire.

XII.

We consumed the forests, the living one above the ground

and the dead ones below. We resurrected

black blood of long dead animals from subterranean reservoirs

and conferred a parody of breath on them once more

as we gulped it into our furnaces and exhaled it

in great plumes from our chimney stacks

in the extravagant necromancy of our zombie economics,

and the fire in our mind made life unbearable,

and that is what separated us from the animals

to travel alone in our metal ark of extinctions

upon the sea of fire that flooded our mind

in the forty thousand generations of our species.

We weren't orphans, we were in self-imposed solitude

as isolated and erratic as the sparks and embers

lifted by rising air above the fires around which we danced,

which we worshiped, which we guarded as eternal

or rekindled at Olympus, the campfire that keeps the dark at bay

as it prowls, hungry, indiscriminate, waiting to eat us

beyond the light demarcating the provinces of animal and man.

XIII.

We assumed we were superior to animals
and exiled them from our ark,
having liberated ourselves from fear with fire,
having repelled all predators with its mystery,
yet what is life for us now but a single safe moment
attenuated over three score years and ten,
a split second of relief diluted with all the time allowed to us,
while the lives of our inferiors are vivid, vibrant,
with a torrent of all the instantaneous terrors and ecstasies
that we have forsworn,
and all our histories echo down hollow centuries of nothing,
and time itself is measured by the palpitations
of a mechanical heart upon the wrist or mantelpiece
that has beat for us since the day our own hearts stopped.

XIV.

We think there are no miracles.
There is only an absence of miracles,
our perception is that the miraculous is exceptional
rather than ubiquitous.
But what we consider miraculous is normal,
and what we consider normal is a world
in which we have stopped being able to see miracles
because we have illuminated it with our fires,
fires that do not replenish the stock they consume,
fires that clear the world for our plough.

XV.

Before he appeared as man,

our god first appeared to us as fire,

in the desert, a burning bush in the wilderness,

a fire that burned while it consumed not,

and as we make our fires in his image,

we discover we are not gods

for our fires destroy what they touch.

XVI.

This is the end of the age of fire.

Our god no longer needs to announce himself with fire.

We are the authors of our own apocalypse.

The fire is our ego,

or rather, its indignation at the world's indifference,

and it would burn the whole world down

and it makes us the authors of the apocalypse

as the climate turns sour,

unless we learn to embrace our own insignificance.

XVII.

We didn't invent the light. We invented darkness:

the night that was once filled with sounds and stars and wonder

was banished beyond the perimeter set by our campfire,

becoming something to fear, and squint at,

with eyes unadjusted to the dark.

God announces himself with fire

only because we have invented darkness,

and when we try to emulate him,

our own announcements - beacons

lit across the forty thousand generations of the age of fire

- take the form of tower blocks burning in the night,

nuclear detonations turning the desert to glass,

concentration camp chimneys,

the ramparts of glaciers collapsing into the sea,

the desperate peregrinations of animals deprived of habitat,

extinctions, and the irreversible bequest of methane

from thawing permafrost and warming ocean sediments.

-

XVIII.

This is what the fire in our hands has made possible.
It is a material with which to conjure darkness,
until the dark side of the Moon is the only Eden we deserve,
one to which we strenuously try to return
among the crowds cheering the moon-shot launch
named after an ancient Greek sky god.
We landed a man safely on the Moon
and returned him safely to Earth fifty years ago.
Now for the rest of us. What about the rest of us?
What Earth do we have to return to anymore?
As for the fantasies of billionaires today:
you cannot escape the fire in a rocket.

XIX.

If I could go back again, from the end of the age of fire

to that first artificial dawn when fire first was kindled

and carried to the back of the cave, a source of fear and awe

that relegated all other fears to the shadows,

that let us overcome those other fears with its light,

that usurped the commonplace as the source of miracles,

I would tell my ancestors this: you are not orphans.

Our mother is not dead. The fire is not real,

and as its hot, incandescent vapours disperse

it would leave only a name, a word, hanging in the air,

an excuse to treat others as though they were as disposable

as the fuel the fire consumed. I would say there is no excuse,

we are not orphans, the earth is not dead and flammable.

We do not need the fire anymore. The sun and wind

will give us all we need. We need not burn anything ever again.

XX.

And I would say our mother forgives us,

Indeed, she requires no redemption,

because her love was always unconditional,

there are no rules she has set which we have transgressed,

no commandments chiselled on tablets of stone,

no tree the fruit of which she has forbidden to us.

All she ever wanted was to know what we need,

and she leaves it up to us to tell her what that is.

Scream

It should have been a scream,
an endless scream,
a scream that didn't stop
until someone heard it
and did something about it
to make it stop.
It should have been a scream.

Instead it was a silence.
Like in Aldermaston,
where strange individuals
handle dreadful things,
and the alarm is always on,
a dull insistent noise,
like a bored administrator
repeating the same advice
over and over and over and over and over again,
and the time to worry,
the time something has leaked,
or gone critical,
or is about to explode
and contaminate or irradiate us,
is when it stops,

it was a silence.

The birds stopped singing,

but we didn't run for cover.

We didn't find a fallout shelter.

We didn't congratulate ourselves

for the foresight

that saw us build a secret stockpile

to get through this.

We didn't run the bath

to use it as a reservoir of water.

We admired glorious sunsets.

They seem more frequent these days,

don't they? We marvelled at the dawn.

We expressed relief that ants

had not ruined our picnic.

So it turned out in the end

it was a scream,

just not the scream

that should have been,

not a scream

we were able to do anything about,

but a scream much worse than that,

and then

an uninterrupted silence began.

Shout

It's not that I don't care:
haven't you noticed the scene I make
when I do express emotion?
The problem is not that nothing registers
enough to make the needle shift,
but that the signal itself is saturated,
my needle always pressed up hard
against the other end of the dial,
not indifferent, but at full scale deflection,
my feelings cramped in constant spasm,
and I can only make myself heard,
on those rare occasions
when the volume is turned down
enough to let me relax a little,
and even then the needle convulses
at the extreme end of its compass
and the only way I can speak
is at the top of my voice.
Otherwise I must remain silent.

Kingdom come

We are kings in patchwork robes
sitting upon irrelevant thrones
unaware our palaces have become asylums.
The hallucination of state has disintegrated
into ten thousand competing lunacies.
The city fell and was abandoned long ago.
No-one notices. Only wolves and their prey
wander back and forth through the breach.

Courtiers squabble. A cascade of fallacy
swept through our solemn deliberations
like a virulent pathogen infecting our logic.
We second guessed ulterior motives,
made those to whom we attributed them
fair game, and argued with the ghosts
of our own fears and fevers and bad faith.

Frustrated by apologising for our own existence,
we argued by accusation, scavenging for grievances
while our enemies whispered wild suggestions,
until at last we opened our minds by lobotomy
and all courtesy became madness,
and now we sit and rule a realm of memory,
and dream of restoration from rags and dust.

The trials

When lies are so commonplace
one assumes all words are lies
and discourse so degraded
we lose the habit of thought
and truth subordinated to authority
held by sham, self-serving charlatans,
it is then the gods release
the trials with which they test us.

The eighteenth green

Deal done and work won,
glasses raised and partners praised,
dictators backed and backs slapped,
regimes propped up and contracts mopped up,
it was getting late down at the club,
the light was fading, and it was time to hit the road.

"Let's just play one last round of golf" you said,
as the barman wearily hollowed out another glass
with that cloth stump where his hand should be.

It was to be a round unlike any other,
the very last round ever. It was clear to me,
even at the time, as I stood up, and put down
my hastily half-drained glass upon the bar,
staggered down to the first tee, and set up
my ball to start, that this would not end well.

We advanced upon the first fairway through barbed wire,

Not all of the caddies made it through the crossfire,

gas and minefields. The green was very well defended

with bunkers, trenches and gun emplacements.

We stopped on Christmas Day to play football for a while,

then returned to our golf clubs. New caddies were called.

We sang "Katyusha" and our rocket launchers screamed,
and from the second tee we launched our final counterstrike,
watching with satisfaction as the green was scorched
beneath the blinding flash and mushroom cloud
unleashed by our thermonuclear ingenuity. That'll teach 'em.

We drove up the third fairway in a stretch limo, in the back seat,
spilling our cocktails in our anonymous companions' hair,
as their heads bobbed up and down in our laps
- an accidental fruit salad, all cherries and pineapple and pigtails -
and eventually reached the casino, the green laid out with baize,
where we rolled dice in Havana in those crazy days before it fell,
and in Las Vegas, Sun City, Monte Carlo, and Macau.
Broke, we placed mah-jong tiles on the tarmac
in some back alley in Shenzhen
in the cold steaming dawn of the long walk home.

If you remember the fourth you weren't there: some drinking den
where memories of missed putts are sunk in shot glasses
by old and broken men, ill lit by crippled candles cut
to make more ends to burn, casting contorting shadows
of pole-dancers who are no longer there,
a writhing floor show beyond the point of no return.

The fifth fairway included a water hazard,

but we managed to play through dry-shod,

stroking our balls from one end of the deck to the other

in a nuclear submarine parked beneath the polar icecap,

being careful not to slice the ball

in case we sent it ricocheting over control panels

and accidentally started another war,

while high-definition polar bears above us despaired

in documentaries watched in well heated homes

around our globally warming world,

their starving cubs receding out of view on the retreating sea ice,

and inside the sub agents, codename Nemo and Noah, held our clubs,

while our Scotch rocks chimed and melted in the glass

beneath the beaming smile of an indoor nuclear sun,

and we toasted all our past victories,

our fortunes now declining in defeat,

and absent friends,

and as the petals fall from the rose I gave you long ago,

my lost but not forgotten love, I commemorate

our private grief here with public catastrophe

and fireballs bloom all around our garden of regrets.

We read about the sixth in Gibbon's Decline and Fall,
in a footnote about Caracalla, one not found in popular abridgements,
noting an apocryphal description of an impromptu detour
while campaigning in Caledonia,
his army deviating from scorched earth and genocide for a few days,
straying off one course to try another in the East Neuk.
No A9 back then. The Romans would have dualled it.
Modern scholarship disputes whether there even is a sixth green.
The Mayan Calendar describes an approach, somewhat cryptically,
from which most stray into the rough, somewhat elliptically ...

And so we ended up stuck in a bunker on the seventh fairway.
We hacked away in a leisurely way in Ravenna and Capri
with waning enthusiasm, occasionally catching the attention
of some idle or imprisoned chronicler whose Consolation
or Secret History we'd read, laugh at and burn,
and on the beach in Ibiza in the afternoon
we'd try our best without success for several centuries
to summon up the effort,
distracted by empire and indolence in equal measure,
until at last Attila said we had incurred a penalty
and told us we should play on, but only once the needle lifted
off the Café del Mar track he was playing.

The hole on the eighth green had been made by a rabbit
and we all know what that leads to.

The ninth green was observed rather than played.
We viewed it calmly, head propped on a forearm,
elbow in one palm, chin in the other,
after a walk through the Tuileries Gardens to the Louvre,
where the picture was hung: the Virgin of the Rocks.
Leonardo painted that secret golf course long ago, hidden,
unseen, in a background landscape behind some angel
and our chubby saviour. We discussed it later in the consulate
over canapés, and were very impressed with ourselves,
until one of us was murdered in mysterious circumstances.
It remains an enduring enigma to this day.
It is a problem to be solved by the hero of a cheap thriller
in an airport lounge during a particularly protracted layover
and everyone that matters is implicated in this conspiracy
that goes all the way to the top of the departures board.

A blind child holds the tenth green up in the palm of his hand,

a pocketful of bleached radioactive dust. He lifts up generations

from the brown ground bone sand of some post-apocalyptic beach,

he holds it out to us, then steps back, forever just out of reach,

leading us on to sink into a thick sick swaying slick of pink poisoned sea

in that half-remembered nightmare that fills our morning with unease,

as we struggle to work out if it was guilt

or premonition that disturbed our sleep

when we crossed the dateline at fifty thousand feet,

waking as the stewardess offers us a cooked or continental breakfast.

We ride a distant vapour trail high in the boy's blue sky

from which the bomb fell on his atoll all those years ago.

Touchdown. The lush, Edenic eleventh was a surprise: a paradise
lost in the jungle since the dawn of time. Hidden from malice.
Life in all its wasteful abundance erupting in sound and colour,
sleeping loudly beneath the moon-spun silken shawl of night,
and rioting gloriously awake in the sun-torn day. In all innocence
we used machetes and chainsaws to hack through the noise
and heat and colour and light and glory of God to get to the green,
at any cost, and illegal loggers attended the flagstick for us.
It took the form of the last hardwood tree,
but they found to their dismay they couldn't put it back,
and so they made expensive furniture out of it instead,
so we could relax in what seemed like luxury at the time.
I'm sure my five-iron hissed an angry complaint at some point
before slipping from my hand and slithering off like a snake,
never to be seen again among the tree stumps
and bare sterile earth eroding in blood red rivulets
to choke the delta with silt.

The twelfth was strip-lit and duty free.
The thirteenth lined with statues.
The fourteenth green was on the Moon.
The Russians wanted to play through,
but we said we would be finished soon
and then lost our clubs on re-entry,
along with the crew.

The fifteenth fairway was patrolled by prostitutes
repeating every token promise we have ever made,
our broken promises and unspoken promises,
our shame, our fame, our blame,
believing our every boast for us. We're all the same:
our staggering swaggering sad bravado,
our offer of a cheap, opportunistic afterthought of flowers,
the whole repertoire of condescension and menace,
until, in a moment of clarity, we realise we are doomed
as a single kiss opens a trap door in our heart,
transforms our stomach into a tar pit
where love eventually goes extinct,
and we sink trying to hold onto each other desperately
with our pathetic wee T-Rex arms flailing
so we don't die alone ...

We fracked the sixteenth.

The seventeenth fairway blew away beneath us

as we made our slow progress on foot - soil erosion,

deforestation, desertification, climate change, whatever, who cares -

until the whole thing was one big bunker.

We found Pompeii and Skara Brae hidden in the dunes

just behind a Manhattan familiar from various disaster movie posters.

You Maniacs! You blew it up! Ah, damn you! God damn you all to hell!

John Knox fell off the Swilcan bridge and drowned,

and all the world's cults perished all around,

in a puddle of privatised water, face down,

and so we walked towards the eighteenth godless and alone.

This eighteenth fairway was built on exhausted urban land
on which was grown the world's last scraps of food
in a future that had always been inevitable, even from the start.
We feigned disorientation as we wandered in our wilderness
of entirely predictable predicaments.
We fertilised that land with what remained of our raw crude,
and on the eighteenth green, where the flagstick stood,
we built the world's last oil rig derrick to get it.

With one stroke we sank the final putt.

We watched the ball roll and fall and rattle in the cup.

We walked across the last few blades of grass,

which we crushed under our feet as we passed

to retrieve the ball from where it came to rest,

only to find it had fallen into a bottomless pit

growing ever wider, the landfill into which we have tipped

everything ever bought,

but which no amount of ash and trash could ever fill,

a pit whose edges crumble under the weight of malls and parking lots,

a pit into which everything we ever built collapses.

We sought collective salvation from our individual sins

Folding hands in prayer, folding paper as money,

but the toppling steeples and minarets,

the fallen temples and trade centers still feed it

as the world buckles, collateral damage

in a war we wage only on ourselves.

You finally turned to me and with a smile said "that was good.
Wasn't it worth it? Not a waste of a good walk.
What do you think?"

and as the last stunted uncut harvest burned down all around,
its thin tinder dry stems planted in fracked and gas-soaked ground,
and all the bushes blazed, not with announcements of divinity,
but with that single quiet chorus of mute doom
(the void commemorates our passing only with unending silence),
and as the sparks and embers rose and mingled with the stars,
I said "it certainly wasn't a waste of a walk, but I need a drink"
and we returned directly to the bar.

Time

Our messiahs betray us.
We endure their betrayal
by imagining we betray them.

We worship our ancestors
and the future is a palace
we imagine for their ghosts.

The first people to possess anything
were the dead, the first possessions
the objects we placed in their graves,

and all the living have owned ever since
was the duty to care for their possessions,
our palaces mausoleums of the imagination.

Responsibility passed from father to son
and so the first wealth was inherited,
and poverty was invented to preserve it.

There is no eternity, there is only now,
the pathways energy takes through the world,
of which each life goes only part of the way,

but we imagine time exists so we can place
objects in eternity, gifts for the dead,
and nothing belongs to the world anymore,

and it becomes disposable, and is discarded,
and the world becomes a dead end, a dump
on which we raise our Calvaries,

and all our messiahs betray us,
and we endure their betrayal
by imagining we betray them.

Tired

I'm tired. So tired.
It's not just fatigue anymore,
not something sleep can fix,
it's existential exhaustion.
I can't imagine not being tired.
I don't care any more
about any achievements
on which my reputation rests.
I can't be bothered
defending a legacy
it took a lifetime to build.
Let the spivs and charlatans
have their way. I surrender.
I'm just too tired.

14

I had 14 years left to live.

I lived them. Then

there were 14 weeks,

14 days, 14 hours. Now

there are 14 minutes left,

and all I'm doing is waiting

until there are only 14 seconds.

I'm leaving it until

the very last moment you see.

I'll say what I have to say

only when there's no other option,

when it can't be put off any longer.

14 seconds might be enough to say it.

They might not. Time will tell.

Sloth

Back off.
Leave me alone.
I'm not being lazy.
This is more
than mere indolence,
or less than:
laziness is effort
compared to this.

It's not that
I'm disinclined
to lift a finger.
Some limitless inertia
rests on me
and makes
all motion impossible.

My heart does not
circulate blood,
but siphons it
somewhere else,
an otherwise
inaccessible chamber
whose capacity is so vast
my few drops are

a vanishingly infinitesimal
addition to the void
pumped into a firmament
filled with black stars
among whose constellations
my own heart is lost.

I do not just sit still.
I imagine myself
draining away,
becoming empty,
like a cavern
leeched from limestone
by centuries of rain
to provide a venue
for the futile echoes
of stalactites
weeping in the dark.

Some things scar,
some stay scabs,
the damage never undone,
and we never heal
to fill the hole
we had to learn
to grow around.

Jump cut

This is a new feeling:
not love, but grief.
The one I love lies dead.

I struggle to express myself.
I handle his tools: his bow and quiver;
the axe-head he knapped from flint,
kneeling by my side in the cave mouth
while the fire cooked what we caught
together; the agate necklace he wore.

His possessions have a new meaning
now that he's dead. They serve a new end.
They are something of him that lasts
and will carry a sense of him
that he himself can no longer bear.

I try to press this new meaning
onto the old words we once used
merely to talk about here and now
and what to do next
so I can describe a better place
where he can still live,
even after I too have gone,
and can no longer sing his exploits,

a place among the ancestors
who were like him once,
where we all must go some day.

These new meanings of old words
turn the objects they describe
into possessions fit for the ancestors,
so nothing belongs to this world anymore.

We dig a grave and lay his body in it,
and place his possessions with him,
and speak to the ancestors,
saying he is worthy of their company
and asking them to receive him,

but they do not hear us. What we do not know
is that we do not speak to our ancestors.
We speak to our descendants,

and not just because they are the ones
who will at some point open this grave
and find the artefacts we placed in it
rather than ancestors who are, after all, dead,

but because this new way of speaking
by the side of a hole in the ground

lets them make the tools they need
to build cities in the sky.

In setting things aside
we make possessions of them,
so that nothing in the world
will ever be the same again
and by imagining a past
that ancestors can inhabit
we have filled the future
with infinite possibilities,

and the grave we dig, all the graves,
in which we lay forty thousand generations
between now and then
collapse together to form one single grave
through which we burrow to the future
where our descendants build a launch pad
and the flames of a Saturn V rocket flare,
pouring a libation of fire upon the past,
as we embark upon a journey
not to the underworld,
but to the stars.

The well

As boys we sat at the edge of the village well
and flicked pebbles into the cool shadow of its shaft
waiting to hear them reach the bottom. We waited.

We waited. The sun shuttled between the solstices
and Eratosthenes tossed those in the well at Syene too
at the height of Summer. Time passed. No sound,

and as we waited, in the full mirage of our youth
the well slowly filled with the shadows of all the words
we did not say, and every pebble became a poem,

falling, and we waited. A butterfly was the pinkie hinge
upon which all our years' vagrant seasons swung,
a miracle of reckless yet enduring delicacy,

of weightless fragility. Early autumn held its breath,
teaching us in silence what cannot be learned
from all the world's loud boasts of strength,

and in the evening the equinox stood at the door,
pausing, searching through his bundle of keys
for the one which unlocks winter, and turned it.

Lightning briefly cracked the sky apart
and for a moment, through the crack, we caught a glimpse
of the light of better worlds, where butterflies are from,

and imagined our pebbles, falling, and as they fell,
growing, accreting mass, becoming planets lost to an orbit
through darkness deeper than any well, and we waited,

and imagined ourselves growing older, accumulating years
in silence, and when we died we imagined ourselves buried
in the cold earth of those planets, falling through space,

and now it is our turn to fall, through the cold grave
and into the bottomless darkness, and as we fall
through the silence of all the words we never said

we realise the only thing we have when all else is discarded
is the fragile love between us, as we wait in the sunshine,
wait in the full mirage of our youth, in silence, by the well.

Eden

The planet is not dying.
In the landfill and debris
we are refounding Eden
one way or another.
The Tree of Life will grow,
whether we cultivate it
or it erupts as a weed
through the dead knuckles
we once clenched tight
around the world's throat.

Our material progress is achieved
in defiance of God, whose glory
should be our sole concern,
and sin, not deprivation, the only way
to explain our misery: therefore,
His word will be heard again,
as a whisper among the atoms
swirling in the wake of our passing,
a rumour infiltrating the world's ruin,

and that rumour is:
there only ever was one Eden,
and it is eternal, it is without end.
Certainly, there was a fall from grace,

and that which fell will rise again,
but it is only a conceit of Man
that it is we who fell and will ascend,
at least not unless everything we are

is entirely undone, and the rumour
whispered among the atoms reveals
all human history was a mirage,
a delirium, an hallucination,
a fever that has run its course,
and the world is now restored
to health among the rubble
which we think represents disaster
only because all the meticulous order
we thought brought to the world
was just a symptom of the chaos
we inflicted upon it. So, Civilisation

is coming to an end? No.
Our institutions may be a circus,
its ringmasters clowns, perhaps,
unfit for purpose, but as we run
our Civilisation into the ground,
and plough its cities under once again,
we are refounding Eden. Our mistake
was ever to imagine it was a garden.

Decade

Time's up.
The masks have all slipped.
All evils are revealed,
smirking,
strutting openly among us
with spurious entitlement,
striding through a shattered fourth wall
and into the stalls, knee deep in punter,
proud of their egregious, indulgent trespass
upon sanity and reason.
We can no longer pretend.
They openly dare us to defeat them.

We can no longer ignore
the questions we must ask.
We can no longer tolerate petty foibles
of wealth and privilege,
now that these combine to kill us.
Eccentricity becomes aggression,
the quaint becomes a cunt,
the toff becomes a tool,
and the indifference in the world
has reached toxic levels of concentration.
Nothing is without consequence anymore.

The Earth is an ark of the damned
adrift upon a sea of fire
and we have ten years to save it.
Otherwise life is left to wither and perish
while all the wealth accumulated
over preceding millennia of civilisation,
all the regalia stolen from the Incas,
each diamond plucked from the gravel of Serendip,
every gold tooth pulled from a mouth in Auschwitz
to be turned into bullion hidden in a vault in Zurich
or London or New York,
is invested in escape routes
for those who brought this disaster on us,

this cataclysm that has overtaken us like a tsunami,
like a hurricane, like a famine, like a drought,
like a flood, like an earthquake, like a war, like a plague,
from which they imagine they are inoculated
by some magic serum distilled from their gold
and injected and smoked and swallowed and snorted
while others starve and suffocate and drown
and are caught in a crossfire of catastrophes.

But it is also a decade of radical hope,
when to resist despair becomes a revolutionary sentiment,
to refuse to give up is a rejection of impossible odds
stacked against you for centuries,
to roll dice loaded by a history of privilege and prejudice,
in defiance of our lop-sided fate.

It is the decade when the men who went to the Moon
truly return to home and find out what has become of it.
The decade when we can build a lasting future,
and finally launch the planet as Spaceship Earth,
and all be in the same boat at last.

For Siddhu

Let us stand together and watch the last sun set
upon our futile barricade, and know we stood
for a while at least against the bitter century
that rises with that bitumen dawn against which
the last embers of our souls burned in vain.

Ashes

You see those stars?

That's the world burning down

in ten thousand little bonfires.

That's how much I love you:

I don't have time

to show you all the ways

- I'd need to be immortal -

so instead, I set fire to the world

to light the way to world's end.

I'll be gone, but my love will endure,

and you will find it waiting there for you,

my wedding ring among the world's ashes.

Can of worms

As passing bells resound the knell
of happier times, I'd like to tell
a tale of worms and the man
who charmed them back into the can.

He played a tune upon a flute
to soothe the wee, unruly brutes,
and when they were back in the tin
said "put this can straight in the bin".

The fate of the poets

Scene: A midnight campfire in a post-apocalyptic wasteland. A few malnourished survivors wracked with radiation sickness gather to cement the fragile relationships that constitute what remains of society. Having exhausted all noteworthy items of information worth discussing, they grow quiet. Then one of them interrupts the silence with a convivial outburst of enthusiasm:

"Here's to the Ballads of Billy the Bard!

His poetry rhymed and it wasn't too hard

to get what he meant when he spun us a yarn,

so we clothed him and fed him and kept him from harm.

"But nobody liked Johnny Troubadour's poems.

No-one could tell where his stories were going:

obscure, indecipherable, complete waste of breath -

so we starved all his family and clubbed him to death."

Acknowledgements

"Solstice" first appeared in Winter Solstice Anthology co-edited by Marie Lightman and Richard Skinner
https://wintersolsticeanthology.wordpress.com/

"The well" first appeared in Causeway / Cabhsair: A Magazine of Irish and Scottish Writing, Volume 8, Issue 2 (2017)

Peter J M Clive

Peter lives on the southside of Glasgow, Scotland with his wife and their three children. He is a scientist who has worked in the renewable energy sector for nearly two decades. As well as poetry, he enjoys composing music for the piano and spending time in the Isle of Lewis and St Andrews with family.